PLOT SUMMARY

"Tomorrow Never Dies," a 1997 James Bond film, follows the British secret agent 007, portrayed by Pierce Brosnan, as he battles media mogul Elliot Carver, who seeks to incite global conflict through sensationalist news reporting. Carver's plan involves manipulating world events and triggering a war between major powers to increase his media empire's profits. Bond teams up with a Chinese secret agent, Wai Lin, to thwart Carver's scheme, navigating a series of high-stakes action sequences and uncovering the extent of the mogul's conspiracy. The film blends espionage with a critical look at media influence and power.

In the article about "Tomorrow Never Dies," the following characters are highlighted:

1. **James Bond (Pierce Brosnan)** - The British secret agent tasked with stopping the villainous plot. Bond is known for his charm, resourcefulness, and expertise in espionage.

2. **Elliot Carver (Jonathan Pryce)** - The primary antagonist, a powerful media mogul who plans to incite global conflict to boost his media empire. Carver is manipulative and ruthless.

3. **Wai Lin (Michelle Yeoh)** - A Chinese intelligence agent who collaborates with Bond to thwart Carver's scheme. She is skilled in martial arts and a key ally in the mission.

4. **Paris Carver (Teri Hatcher)** - Elliot Carver's estranged wife, who has a past romantic connection with Bond. Her involvement adds personal stakes to the mission.

5. **Admiral Roebuck (Geoffrey Palmer)** - A British naval officer who becomes embroiled in the conflict as Carver's plot threatens international relations.

These characters drive the film's plot, each contributing to the dynamic interplay of espionage and action.

THE SETTINGS

The settings in "Tomorrow Never Dies" include:

1. **Shanghai, China** - The bustling metropolis where much of the film's action takes place, featuring high-tech environments, luxurious settings, and intense chase sequences.

2. **The Stealth Ship** - Elliot Carver's advanced, stealth-equipped ship, which plays a central role in his plot to incite global conflict.

3. **The Carver Media Headquarters** - A sleek, high-tech media building in Hamburg, Germany, serving as Carver's base of operations and a key location for the film's climax.

4. **Vietnam** - The tropical backdrop for dramatic action scenes, including Bond's encounters with Carver's henchmen and significant plot developments.

5. **London, England** - Briefly featured as the location where Bond receives his mission and reports back to MI6.

These varied settings contribute to the film's global scope and dynamic action sequences, highlighting the international stakes of the conflict.

THE THEME

The theme of "Tomorrow Never Dies" centers on the influence and power of media in shaping global events. The film explores how a media mogul, Elliot Carver, manipulates news and information to provoke international conflict and achieve personal gain. It critiques the ethical boundaries of media, highlighting the potential dangers of sensationalism and the impact of unchecked power on global stability. The story underscores the importance of truth and integrity in journalism while showcasing the role of espionage in countering such threats.

TOMORROW NEVER DIES

CHAPTER ONE

In the quiet, early hours of dawn, the world outside seemed to hold its breath. The horizon was a soft blend of twilight hues, promising a new day that would unfold in countless, yet familiar, ways. The notion of "Tomorrow Never Dies" is not merely a poetic reflection; it is a profound commentary on how we experience time and our place within it.

In this first chapter, we embark on a journey through the intricacies of our daily lives, examining how the concept of an ever-present tomorrow influences our decisions, actions, and perceptions. Time, often regarded as an unyielding force, becomes a central character in our narrative—a character that we cannot escape, but with whom we have a complex relationship.

The phrase "Tomorrow Never Dies" suggests that tomorrow, in its essence, is an ever-receding horizon. It is a promise of future possibilities, an opportunity to start anew, and yet it remains perpetually just out of reach. This elusive quality of tomorrow drives human behavior, pushing us to strive for a better future while simultaneously causing us to defer our present actions.

Consider the way we often approach our goals and dreams. The promise of tomorrow provides comfort, allowing us to postpone our ambitions with the assurance that we will have another chance. This deferral can be a double-edged sword. While it offers the hope of improvement and growth, it can also lead to procrastination and missed opportunities.

Our decisions are heavily influenced by our perception of time. We often weigh the immediate discomfort of action against the distant benefits of procrastination. This cognitive bias, known as temporal discounting, skews our judgment, causing us to prioritize short-term relief over long-term gain.

Take, for example, the common resolution to start a new habit or make a significant life change. Many of us set goals with the intention of beginning "tomorrow," only to find ourselves caught in a cycle of inaction. This is where the notion of "Tomorrow Never Dies" plays a crucial role. By continually deferring our actions, we create a feedback loop where the promise of tomorrow becomes a barrier to progress.

To counteract the effects of procrastination, we must learn to embrace the present moment. The key lies in understanding that while tomorrow is always a part of our planning, the present is the only time we truly have control over. By focusing on what we can achieve today, we break free from the illusion that tomorrow will magically resolve our issues or bring us closer to our goals.

Living in the moment requires a shift in perspective. It involves recognizing the value of our current actions and their immediate impact on our lives. This approach fosters a sense of urgency and purpose, driving us to make meaningful changes now rather than later.

As we move forward in this exploration, we will delve deeper into how the concept of an unending tomorrow shapes our lives. We will examine personal anecdotes, psychological theories, and practical strategies to harness the power of today. By understanding the interplay between our present actions and the promise of tomorrow, we can better navigate the complexities of our existence and achieve a more fulfilling life.

In the next chapter, we will explore the psychological and emotional aspects of time perception, investigating how our relationship with tomorrow influences our mental well-being and overall life satisfaction.

In this opening chapter, we have laid the groundwork for a comprehensive examination of how tomorrow, despite its perpetuity, shapes our lives and choices. As we continue, we will unravel the deeper layers of this concept, seeking insights that will illuminate our journey through the ever-

Comprehensive Examination of "Tomorrow Never Dies"

"Tomorrow Never Dies," a 1997 James Bond film, provides a nuanced exploration of media influence and its potential for manipulation. Set against a backdrop of global intrigue, the film features a media mogul, Elliot Carver, whose ambition drives him to orchestrate international conflict to expand his empire.

The film's plot delves into the ethical implications of media sensationalism, showcasing Carver's use of false reporting to incite war. Through high-stakes action sequences and complex character interactions, "Tomorrow Never Dies" critiques the power of media to shape and distort reality. By examining this dynamic, the film underscores the importance of media integrity and the consequences of its abuse on global affairs.

TOMORROW NEVER DIES

CHAPTER TWO

As dawn turns to day, we begin to unravel the psychological and emotional dimensions of our relationship with time. Chapter Two of "Tomorrow Never Dies" delves into how our perception of an infinite future influences our mental well-being, decision-making processes, and emotional responses.

At the heart of our exploration is procrastination—a common yet complex behavior rooted in our relationship with time. Procrastination often stems from an over-reliance on the promise of tomorrow, leading us to delay tasks and responsibilities. Psychologists suggest that this tendency is driven by a combination of fear, discomfort, and a lack of immediate accountability.

When faced with tasks we find unpleasant or overwhelming, the notion of a distant tomorrow offers a temporary escape. However, this delay frequently leads to increased stress and anxiety as deadlines approach. Understanding the psychology behind procrastination helps us to recognize the internal conflicts at play and address them more effectively.

Our ongoing deferral of actions and decisions can significantly impact our mental health. The constant anticipation of a better future while neglecting the present can lead to feelings of inadequacy, frustration, and depression. This emotional strain is often compounded by the awareness of missed opportunities and unfulfilled potential.

Research shows that individuals who habitually postpone tasks or decisions may experience heightened levels of anxiety and lower overall life satisfaction. By learning to manage our expectations and embrace the present moment, we can mitigate some of these negative effects and foster a more positive outlook.

While procrastination is a challenge, the anticipation of tomorrow also has a constructive role. Planning for the future allows us to set goals, envision possibilities, and prepare for challenges. This forward-thinking approach can be highly motivating and provide a sense of direction.

Effective planning involves setting realistic deadlines, breaking tasks into manageable steps, and maintaining a balance between immediate and long-term goals. By integrating these strategies, we can harness the benefits of anticipation while avoiding the pitfalls of procrastination.

To counteract the tendency to defer action, we must develop strategies to embrace the present. This involves cultivating mindfulness and developing a proactive attitude. Mindfulness techniques, such as meditation and reflection, can help us stay grounded in the current moment, reducing the allure of procrastination.

Additionally, setting specific, achievable goals and celebrating small victories can create a sense of accomplishment and reinforce positive behavior. By focusing on what can be achieved today, we create momentum that propels us toward our larger objectives.

Striking a balance between living in the moment and planning for the future is crucial. While it is essential to address immediate responsibilities and enjoy the present, we must also remain mindful of our long-term aspirations. This equilibrium allows us to navigate the complexities of time without succumbing to the pitfalls of procrastination or losing sight of our future goals.

As we progress to the next chapter, we will explore how different cultures and philosophies interpret the concept of time and its influence on our lives. By understanding diverse perspectives, we can gain a richer appreciation of how our relationship with time shapes our experiences and actions.

Chapter Two provides insight into the psychological and emotional dimensions of our interaction with time. By addressing procrastination, mental health impacts, and strategies for embracing the present, we lay the foundation for a deeper understanding of how our perception of tomorrow affects our lives. As we continue our journey, we will explore various cultural and philosophical perspectives on time, further illuminating the intricate dance between our present and future.

Tomorrow Never Dies:

Chapter Three

As we delve into Chapter Three of "Tomorrow Never Dies," we shift our focus to how different cultures and philosophies interpret the concept of time and its impact on human life. Understanding these diverse perspectives provides valuable insights into how our relationship with time can shape our experiences and guide our actions.

Time is perceived and valued differently across cultures, and these interpretations profoundly influence behavior and societal norms. In Western cultures, time is often seen as linear—a straight path from past to future with clear demarcations such as deadlines and schedules. This linear perspective fosters a sense of urgency and efficiency but can also lead to stress and the constant pressure of time management.

In contrast, many Eastern cultures view time in a more cyclical manner, where past, present, and future are interconnected in an ongoing loop. This cyclical perspective emphasizes harmony and continuity, encouraging individuals to focus on balance and the present moment rather than the relentless pursuit of future goals.

Philosophers throughout history have offered various interpretations of time, each contributing to our understanding of its role in human life. Ancient Greek philosophers like Heraclitus posited that time is a constant flux, where everything is in a state of change and impermanence. This view suggests that clinging to a fixed sense of time can lead to frustration, as change is the only constant.

Conversely, Immanuel Kant proposed that time is a mental construct—a framework through which we perceive and organize our experiences. According to Kant, time does not exist independently of our perception; rather, it is a part of our cognitive processes that helps us make sense of our existence.

Cultural attitudes towards time shape various practices and traditions. For instance, in cultures with a strong emphasis on punctuality and deadlines, such as in many Western societies, there is often a high value placed on efficiency and productivity. This can lead to rigorous schedules and a focus on future outcomes.

Conversely, in cultures with a more relaxed attitude towards time, such as some Indigenous cultures, there is often a greater emphasis on the present moment and community well-being. These cultures may prioritize relationships and experiences over strict adherence to time constraints, fostering a more holistic and less stressful approach to life.

In our contemporary, globalized world, the blending of cultural perspectives on time has led to an increased emphasis on time management. With the rise of digital technology and the fast pace of modern life, individuals are constantly juggling multiple demands and trying to optimize their use of time.

This modern quest for efficiency often reflects a Western-oriented approach to time, where productivity and achievement are highly valued. However, integrating elements from other cultural perspectives—such as mindfulness and balance—can offer a more nuanced approach to managing time. Embracing these diverse strategies can help individuals achieve a more harmonious and fulfilling life.

As we reflect on the various interpretations of time, it becomes clear that our relationship with it is deeply personal and influenced by cultural, philosophical, and practical factors. By exploring these perspectives, we gain a broader understanding of how time shapes our behavior, values, and experiences.

In the next chapter, we will examine the impact of time on personal identity and life purpose. We will explore how our perception of time influences our sense of self, our long-term goals, and our overall sense of fulfillment.

Chapter Three sheds light on the cultural and philosophical dimensions of time, revealing how different interpretations influence our lives and practices. By understanding these diverse perspectives, we can better navigate the complexities of our relationship with time and find a balance that aligns with our values and aspirations. As we move forward, we will continue to explore how time affects our personal identity and sense of purpose, further enriching our understanding of this fundamental aspect of human experience.

Tomorrow Never Dies:

Chapter Four

In Chapter Four of "Tomorrow Never Dies," we turn our attention to the intricate relationship between time, personal identity, and life purpose. This chapter explores how our understanding of time influences our sense of self, our aspirations, and our overall sense of meaning in life.

Our personal identity is deeply intertwined with our perception of time. From a young age, we begin to construct our sense of self through the experiences and milestones we encounter. These experiences are often framed within a temporal context: childhood, adolescence, adulthood, and old age. As we navigate these stages, our identity evolves, influenced by how we perceive and relate to time.

For instance, significant life events such as graduations, career changes, and relationships can redefine our sense of self. The way we interpret these events—whether as opportunities, challenges, or markers of progress—shapes our personal narrative and identity. Our understanding of the past and anticipation of the future play crucial roles in this ongoing process.

Our goals and aspirations are also profoundly influenced by our perception of time. When we set long-term objectives, such as career aspirations or personal achievements, we often project these goals into the future. This forward-thinking approach can be motivating, providing direction and purpose. However, it also requires us to balance these aspirations with the realities of the present.

The pressure to achieve future goals can sometimes overshadow the importance of living fully in the present. The challenge lies in finding a harmonious balance between striving for future success and appreciating current experiences. By setting realistic milestones and celebrating progress, we can align our aspirations with our present actions, creating a more integrated and fulfilling path forward.

The concept of life purpose is intricately linked to our perception of time. Many people seek meaning through their work, relationships, and personal achievements, striving to make a lasting impact. This quest for purpose often involves contemplating our legacy and the ways in which our actions contribute to a greater whole.

Our sense of purpose can be influenced by how we view time. For example, those who see time as a finite resource may prioritize activities that offer immediate gratification or significant personal impact. Conversely, those with a more cyclical or expansive view of time may focus on ongoing contributions and experiences that align with their values.

To navigate the complex interplay between time, identity, and purpose, it is essential to cultivate self-awareness and flexibility. Regular reflection on our goals, values, and experiences can help us align our actions with our evolving sense of self. Practicing mindfulness and staying present can also enhance our ability to appreciate and make the most of the current moment while working towards future aspirations.

Additionally, fostering resilience in the face of setbacks and adapting our goals based on changing circumstances can contribute to a more balanced and fulfilling approach to life. By embracing the dynamic nature of time and its impact on our identity and purpose, we can create a more harmonious and meaningful life journey.

As we proceed to the next chapter, we will explore the practical aspects of time management and how effective strategies can enhance our productivity and well-being. We will examine techniques for optimizing our use of time, setting achievable goals, and maintaining a healthy work-life balance.

Chapter Four examines the relationship between time, personal identity, and life purpose, highlighting how our perception of time shapes our goals, aspirations, and sense of meaning. By understanding and navigating this relationship, we can create a more integrated and fulfilling life path. In the next chapter, we will focus on practical strategies for managing time effectively, providing tools and insights to help you enhance productivity and achieve a balanced life.

Tomorrow Never Dies

chapter five

"Tomorrow Never Dies" picks up the tension as James Bond continues to unravel the machinations behind the film's central plot. The chapter delves deeper into the nefarious schemes of the antagonist, Elliot Carver, and Bond's ongoing investigation into Carver's media empire.

The chapter is set against the backdrop of Carver's vast media empire, which Bond suspects is more than just a commercial enterprise. The narrative unfolds with Bond in a high-stakes environment, where he attempts to understand Carver's ultimate goal and how it ties into the broader geopolitical tensions.

- Bond investigates Carver's operations with the aid of his MI6 contacts. The chapter reveals more about the sophisticated and invasive media control exerted by Carver. Bond uncovers evidence suggesting that Carver is not only manipulating news but is also involved in illicit activities that could have global repercussions.

- Bond meets with his allies to discuss the information he has gathered. This includes his interactions with M, who provides guidance and additional resources for the investigation. Bond's relationship with M is highlighted, showcasing their mutual respect and the critical role M plays in strategizing Bond's approach.

  - Bond begins to piece together Carver's plan to incite global conflict through his media network. The chapter details Bond's realization that Carver is orchestrating events to manufacture news that will lead to international instability, which he intends to exploit for financial gain.

  - The chapter includes tense action sequences where Bond confronts Carver's henchmen. These scenes emphasize Bond's resourcefulness and combat skills, showcasing his ability to handle dangerous situations while furthering his investigation.

  - Bond uncovers more about Carver's personal motivations and his background. The chapter explores Carver's ambition and his twisted vision of world power through media dominance. This deeper insight into Carver's psyche helps Bond and the reader understand the scale of the threat.

- **James Bond:** Bond is portrayed as both a skilled operative and a determined investigator. His commitment to uncovering the truth and stopping Carver's plan is evident throughout the chapter.

- **Elliot Carver:** Carver emerges as a complex villain with a grandiose vision. His media empire is revealed to be a facade for more sinister goals, adding layers to his character and increasing the stakes for Bond.

- **Supporting Characters:** The interactions between Bond and his allies, including M and other MI6 operatives, are crucial in advancing the plot. These relationships help develop Bond's character and highlight the collaborative nature of his mission.

- **Media Manipulation:** The chapter reinforces the theme of media manipulation and its potential to influence global events. Carver's use of media to create conflicts reflects real-world concerns about the power of information and misinformation.

- **Power and Ambition:** Carver's ambition to control global narratives through his media empire underscores the theme of power and the lengths to which individuals will go to achieve their goals.

Chapter 5 of *"Tomorrow Never Dies"* is pivotal in advancing the plot and deepening the reader's understanding of the central conflict. Bond's investigation into Carver's operations reveals the scope of the threat and sets the stage for the subsequent action. The chapter skillfully combines elements of suspense, action, and character development, maintaining the momentum of the story and heightening the anticipation for the unfolding drama.

Factors to Consider for Tomorrow:

A Comprehensive Guide

As we look toward the future, planning and decision-making become crucial in navigating an increasingly complex world. Whether you're considering personal goals, professional endeavors, or broader societal issues, several factors should be taken into account to ensure a successful and informed approach to tomorrow. This article explores key factors to consider as you prepare for the future, offering a roadmap to guide thoughtful and strategic planning.

Technological progress continues to reshape every aspect of our lives. Emerging technologies such as artificial intelligence, blockchain, and quantum computing promise to transform industries and create new opportunities. When planning for tomorrow, it's essential to stay informed about these advancements and understand their potential impact on your field or personal life. Consider how you can leverage technology to enhance your skills, improve efficiency, and stay competitive in an evolving landscape.

Economic conditions significantly influence decision-making and planning. Factors such as inflation rates, interest rates, and global trade dynamics affect personal finances, business strategies, and investment decisions. Staying updated on economic forecasts and trends can help you anticipate changes and adapt your plans accordingly. Evaluate how shifts in the economy might impact your financial goals, career prospects, or business operations.

Environmental concerns are becoming increasingly critical as we face challenges such as climate change, resource depletion, and biodiversity loss. Incorporating sustainability into your planning is essential for ensuring long-term viability and minimizing negative impacts. Consider adopting eco-friendly practices, supporting sustainable initiatives, and preparing for regulatory changes related to environmental protection. Understanding your role in promoting sustainability can contribute to a more resilient and responsible future.

Social and cultural dynamics play a significant role in shaping the future. Changes in societal values, demographic shifts, and cultural trends can influence everything from consumer behavior to workplace expectations. Stay aware of emerging social issues and cultural changes that may impact your personal or professional life. Understanding these shifts can help you align your goals with broader societal trends and engage meaningfully with diverse communities.

The importance of health and well-being cannot be overstated when planning for the future. Physical and mental health are fundamental to achieving long-term success and happiness. Consider incorporating wellness practices into your daily routine, such as regular exercise, healthy eating, and stress management. Additionally, staying informed about advancements in healthcare and wellness can help you make proactive decisions regarding your health and lifestyle.

Continuous learning and skill development are crucial for staying relevant and competitive. As industries evolve and new fields emerge, the demand for new skills and knowledge increases. Evaluate your current skill set and identify areas for growth. Consider pursuing further education, professional development, or certifications to enhance your expertise and adapt to changing demands. Investing in your education ensures you are prepared for future opportunities and challenges.

Political and legal environments can have a significant impact on personal and professional decisions. Changes in legislation, government policies, and political stability affect everything from business regulations to individual rights. Stay informed about political developments and legal changes that may impact your plans. Understanding the regulatory landscape can help you navigate potential challenges and take advantage of new opportunities.

Global events and geopolitical risks can have far-reaching effects on various aspects of life. Issues such as international conflicts, pandemics, and economic sanctions can disrupt plans and create uncertainties. Monitor global news and developments to anticipate potential risks and prepare contingency plans. Being aware of geopolitical factors can help you make informed decisions and mitigate potential impacts on your goals.

Effective financial planning is essential for achieving long-term goals and ensuring stability. Consider factors such as savings, investments, and debt management when preparing for the future. Develop a comprehensive financial plan that includes budgeting, retirement savings, and risk management strategies. Regularly review and adjust your financial plan to reflect changes in your circumstances and financial goals.

Aligning your plans with your personal values and long-term goals is crucial for finding fulfillment and success. Reflect on what matters most to you and how your decisions align with your values. Setting clear, achievable goals and developing a plan to reach them can provide direction and motivation. Balancing personal aspirations with practical considerations ensures that your plans are both meaningful and attainable.

Preparing for tomorrow involves a multifaceted approach that considers technological advancements, economic trends, environmental sustainability, social changes, and personal well-being. By staying informed and proactive, you can navigate the complexities of the future and make decisions that support your goals and values. Embracing a holistic perspective and adapting to evolving circumstances will help you build a resilient and successful path forward.

What If Tomorrow Dies?

"Tomorrow Never Dies" offers a compelling exploration of the potential consequences if media influence becomes unchecked and manipulative. The film, set against a backdrop of global intrigue and espionage, poses the question: "What if tomorrow dies?" This question can be dissected both literally and metaphorically, revealing profound implications for our understanding of media power and its effects on society.

1. **Literal Interpretation**: In the literal sense, if "tomorrow" were to die, it would imply an immediate and catastrophic collapse of future possibilities. The film portrays a scenario where the manipulative actions of Elliot Carver, a media mogul with grandiose ambitions, threaten to trigger an all-out war between major powers. Carver's plan involves using sensationalist media coverage and false reporting to create conflicts that could spiral out of control. If his scheme succeeds, the resultant chaos could lead to widespread destruction, undermining global stability and safety. This potential for apocalyptic outcomes highlights the dangers of allowing media power to go unchecked and demonstrates how the pursuit of personal gain can have dire consequences for humanity.

2. **Metaphorical Interpretation**: On a metaphorical level, the idea of "tomorrow dying" represents a loss of hope and future prospects. In the film, Carver's actions symbolize a broader threat to societal progress and values. By distorting reality and manipulating public perception, Carver

undermines the public's trust in media and erodes the foundational principles of truth and accountability. This metaphorical death of "tomorrow" suggests a world where the future is marred by deception and corruption, leading to a loss of ethical standards and a diminished capacity for constructive dialogue and growth.

3. **Media Influence and Responsibility**: The film's portrayal of media as a tool for manipulation raises critical questions about the role of media in shaping reality. Carver's control over his media empire allows him to craft narratives that serve his interests, irrespective of the truth. This highlights the significant responsibility that media outlets and journalists have in maintaining integrity and accuracy. If media platforms are used to mislead and incite conflict, the resulting impact on society can be profound, leading to the erosion of public trust and the potential for real-world consequences.

4. **Impact on Society**: The film underscores the societal implications of media manipulation. If media entities like Carver's were to dominate, they could influence public opinion and policy decisions in ways that prioritize sensationalism over truth. This could lead to a society where misinformation becomes the norm, and critical thinking is undermined. The repercussions of such a media landscape include increased polarization, social unrest, and a general decline in democratic processes.

5. **Ethical Considerations**: The exploration of these themes in "Tomorrow Never Dies" serves as a cautionary tale

about the ethical boundaries of media power. It poses the question of how far individuals and organizations might go to achieve their ends, and the moral responsibility they bear in their influence over public perception. By presenting a scenario where media manipulation threatens global stability, the film encourages viewers to reflect on the importance of media ethics and the need for vigilance against the misuse of media power.

In summary, "Tomorrow Never Dies" offers a multifaceted examination of the potential consequences if media influence were to become uncontrollable and manipulative. By exploring both the literal and metaphorical implications of "tomorrow dying," the film highlights the crucial role of media integrity in safeguarding future stability and societal progress. It serves as a stark reminder of the responsibilities that come with media power and the impact that misinformation can have on the world.

"Tomorrow Never Dies" is a James Bond film released in 1997. The movie features Bond, played by Pierce Brosnan, battling a media mogul who plans to incite global conflict to boost his news empire. The film explores themes of media manipulation, technology, and global politics. The title reflects the idea that the influence and impact of media and news are ongoing and unending.

CONCLUSION

In conclusion, "Tomorrow Never Dies" underscores the perpetual influence of media in shaping global events and public perception. Through its portrayal of a media mogul's manipulative schemes, the film highlights the far-reaching impact of media on society and international relations. The title itself serves as a reminder that the power of news and information continues to evolve and resonate long after each story is told, reflecting the enduring nature of media influence in our world.

www.ingramcontent.com/pod-product-compliance
Lightning Source LLC
Chambersburg PA
CBHW031517210526
45464CB00007B/2952